MOTHERS OF DATA SCIENCE

CONTENTS

INTRODUCTION

Hi! This is Kate Strachnyi and Kristen Kehrer — thank you for your interest in our book! We're pleased to share the personal and professional experiences of eleven mothers in data science.

Though we share our stories, too, this book is not about us. It presents qualitative and quantitative research on the advantages and challenges of being a parent in data science, as well as commentary about the opportunities for and advantages of mothers in data science.

We embarked on this journey a year ago, when we started interviewing inspirational data scientists who also happen to be mothers. We wanted to uncover their stories and learn from their experiences and we can't wait to share our findings with you!

APPROACH FOR INTERVIEWS

We researched and identified our candidates and then talked to these amazing mothers. We are so thankful that so many were willing to participate in the interviews and in the book. We learned that writing a book is a fantastic way to speak with people you consider role models.

You can refer to the appendix for the biographies of the women who generously shared their time and experiences with us. Listed in alphabetical order by first name, we spoke with:

Alice Zhao, mother of two small children and senior data scientist at Metis. She is passionate about teaching and mentoring, and loves using data to tell fun and compelling stories.

Carla Gentry, grandmother, mother of two adult sons and a mathematician/economist. Owner and data scientist at Analytical Solution. She has worked in the field of data science for over 15 years. Carla is ranked among the top 10 "Big Data Pros" to follow on Twitter.

Cathy O'Neil, mother of three sons, is known for authoring the book *Weapons of Math Destruction*. She is an American mathematician and the author of the blog *mathbabe.org*.

Claudia Perlich, mother of a teenage boy, is a senior data scientist at Two Sigma. She has published in more than 50 scientific publications and has a few patents in machine learning.

Deborah Berebichez, mother of two young children, is the first Mexican woman to graduate with a physics Ph.D. from Stanford University. Her work in STEM outreach has been recognized by TED, *The Wall Street Journal*, The American Association for the Advancement of Science, Oprah, CNN and Discovery Channel. She is on a mission to inspire young people to pursue careers in science.

Heather Shapiro, mother of a small child, has a Ph.D. in neuroscience and 10-plus years' experience in research and data. She is passionate about startups, health, wellness, wearables, education and the brain.

Jacqueline Nolis, mother of a toddler son, is a data science consultant and mentor. She is also the co-author of *Build Your Career in Data Science*. She has a unique point of view as a transgender woman who became a parent as a father and is raising her son as a mother.

Natalie Evans Harris, mother of a young daughter, is a sought-after thought leader on the ethical and responsible use of data. For nearly 20 years, she has been advancing the public sector's strategic use of data, including a 16-year career at the National Security Agency (NSA), and 18 months with the Obama administration. She is the co-founder and head of strategic initiatives at Bright Hive.

Olivia Parr-Rud, a grandmother and mother of three adult children, has more than 30 years' experience working with data. She has a gift for bridging the left-brain world of analytics, data management and strategy with the right-brain world of human values and creativity.

ABOUT US

Kate Strachnyi, mother of two small children, founder of Story by Data and the DATAcated Academy, and author of *The Disruptors: Data Science Leaders, and the Journey to Data Scientist*. She hosts the video podcast *Humans of Data Science*.

Kristen Kehrer, mother of two small children, and founder of Data Moves Me. She has more than 10 years of experience in industry-building machine learning solutions to solve business problems. Instructor at UC Berkeley Extension, Master of Science in Statistics.

PARENTS OF DATA SCIENCE SURVEY

In addition to the in-depth interviews, we wanted to incorporate data points from third-party sources about mothers in tech in general. We figured that "mothers in data science" was too narrow. Still, we had difficulty finding data.

After an internet search turned up nothing of value, we reached out to businesses that we thought might have some data on mothers in tech. We came up empty.

Typically, if you can't find something you want, you have to create it yourself, so we conducted the "Parents of Data Science" survey. We surveyed all gender representations to see what the overall attitudes were, how people were affected and their experiences. The survey was shared across our social media channels for two weeks. Friends shared and retweeted the survey. It was really nice to see members of the community helping to spread the word.

After the data was collected, it was transformed and cleaned. A dashboard with the survey data can be found at https://datamovesme.com/data4data-nerds if you'd like to interact with the output. The raw data survey responses can be found along with R code to perform the transformation of the original data into data ready for analysis in this blog article:

https://datamovesme.com/2019/07/14/parents-of-data-science-survey-r-code/

This github link contains the transformed dataset:

https://github.com/KristenKehrer/machine-learning-parsnip

WHAT TO EXPECT FROM THIS BOOK

In this book, we cover the advantages and challenges of being a mom in data. You'll hear first-hand stories about pregnancy, maternity leave and being a mother of toddlers, school-age children and teens. You can also read about the results of the *Parents in Data Science Survey* and some words of advice from the mothers we spoke with.

We sprinkle the comments and stories from the mothers we interviewed throughout this book, but once we heard Carla Gentry's story, we decided to present it intact as a single chapter. Her struggles and accomplishments paint a cohesive picture of mothers in data science and greatly inspired both of us, so we felt it deserved to be a chapter in itself.

We are excited to share this book with you!

— *Kate Strachnyi* **and** *Kristen Kehrer*

CHAPTER 1

MOTHERS & DATA SCIENCE: THE ADVANTAGES

We love working in data science. It's transformed how we live. It has provided flexibility for us in many ways, including working remotely, which we have both experienced part time and full time.

We asked our group of mothers what the advantages of working in data science are, and here is what they said.

— Kate & Kristen

Mothers bring a future perspective to data science

Cathy O'Neil

My experience is that mothers have a better perspective because they have to care about more than just themselves. Simply because they have the perspective and that of their children, they have the vision of the future. And that's already a lot more than a hard-nosed focus on just the key performance indicator (KPI).

The future of data science is either going to go the way of the social credit score in China, where you're manipulating people into behaving well with respect to a repressive regime, or it's going to go the way of, *"Hey, why don't we think longer term and broader and for all the stakeholders involved and not just for our company? How do we think bigger with data?"* And the latter would be exciting.

Women's common sense, wisdom, empathy and thinking on behalf of more than just themselves but also the future generations — and other populations that are not represented in the data science community — is what's going to make data science a better place to work.

Women have natural skills that align with data science

Deborah Berebichez

I'm very lucky to be working in data science because it allows different aspects of my personality to flourish–my quantitative side, my love of public speaking and networking, and my interest to be an evangelizer of complex concepts in a technical field. I enjoy it tremendously.

I think data science is and will continue to be a great place for women because it provides an opportunity to combine three skills that come naturally to many women: programming, mathematical thinking and communicating.

Women tend to think in an organized manner, in an algorithmic way. Basically we're very spatially ordered and because we've had generations where we've been managing households and various different schedules, we think of what comes first, what comes after and what relates to what.

For example, women think like, *"If I go pick up this here, then maybe in that location I should also run that other errand."* That's a very algorithmic way of optimizing, so to speak, your life activities. That comes in quite handy in data science.

Mathematics is also something that can be done as part of a group or in isolation. It's something that is deep, fun, and that lends itself to really interesting results. We care about the world, thus the desire to apply our technical capabilities to solving problems of the world.

The third one is communication. I think overall women can be excellent communicators. I think we like to communicate when we find out something insightful and original; we want everybody to know about it. Because of that, I believe women can enjoy the responsibility of taking quantitative insights and communicating them to the public.

Those three things are not only easily adaptable to our schedules, and our complex lifestyles as moms, but also they are things that allow us to be creative and expand our growth in different directions, whether you want to go more into the engineering, more into the mathematics part of data science, or more into the communication, like I've gone into the popularizing data science concepts and whatnot. You can make your career as intense as you want to by optimizing the role that you'd like to play in the data science ecosystem.

The State of Women in Technology

Although we had difficulty finding studies on parents in data science, there is plenty of information about the state of women in tech.

According to "Women in Tech: The Facts", a report by the National Center for Women & Information Technology (NCWIT), "In 2015, women held 57% of all professional occupations, yet they held only 25% of all computing occupations."

The NCWIT report authors believe that this pattern is especially troubling given ample evidence of the critical benefits diversity brings to innovation, problem-solving, and creativity. Indeed, a solid body of research in computing and in other fields documents the enhanced performance outcomes and benefits brought about by diverse work teams.[1]

— Kate & Kristen

---

1 https://www.elderresearch.com/blog/team-diversity-women-in-data-science

Data science provides flexible work opportunities

Claudia Perlich

My son has come with me to a number of data science events. At times my son would help me advance the slides and sit in on the presentations. I remember four years ago, I had to give a keynote in Sydney, so we just made a trip out of it, and we booked an RV (recreational vehicle).

I dragged my son to the conference and he was super happy because, guess what? The conference was the only place where there was Wi-Fi, and he could watch YouTube. Compared with the back roads of Australia, where not much is moving in terms of Wi-Fi.

Jacqueline Nolis

I work remotely at times. On the one hand, I would say it's nice because my position and a lot of the earlier positions I've had as a data scientist have been flexible. When I put in my hours is really variable, and I think that really helps me with parenting stuff. Like when I have a doctor's appointment for my son , I don't have to worry, and that's been extremely helpful.

But working from home when you have your child who stays at home is extremely difficult. I basically can't work from my real home. Any time I "work from home" I go work at a coffee shop to have the space of just going into the office that I know is calm and quiet.

And so, if I were full-time remote and worked from home, I would then have to go get a WeWork space or one of those co-working spaces where I would be away from my child because it is just not possible for me to do anything productive if I'm within 50 feet of my son. So I think, in that case, it's a fair amount harder.

Parents in Data Science Survey

In our survey results, we found that 46% of moms work remotely at least part of the week, and 13% work remotely full time.

Fathers in data science also benefit from working remotely, as 36% of fathers work remotely at least part of the week, and 10% work remotely full time.

— Kate & Kristen

Data science + motherhood is a self-improvement formula

Heather Shapiro

Becoming a mom has helped me in my career because I feel I'm able to prioritize a lot better. I don't have much patience for lower priority or inefficient things, just because any extra time I have I'd rather be spending with my family. I just want to get things done and go home. I'm a little more ruthless about trying to identify the highest-impact, highest-priority things to do.

I think that's a good thing.

I think it's also made me more patient. And I understand that anyone can have any number of things going on in their lives. I didn't think about those things before.

Deborah Berebichez

There's a misconception that when you have kids you will do every-thing else that you did before you had kids, and then on top of it you will have to find extra time to take care of them. That's not true, of course. You have less time to do things.

But what I didn't realize is that you become much more effective at planning things, at using your time. I am much less picky about the food I eat, the times that I eat, the clothes that I wear. Those things just immediately became a lower priority.

I definitely am lucky that I manage a team of amazing 16 data scientists at Metis. They all go through what we call a passion project for one quarter. For three months during a year they go out and work for a company of their choice while they still get paid by Metis, or they write a chapter for a book, or they do whatever they're passionate about having to do with data science. I approve those projects. I read them. I work with them to create thought leadership activities and blog pieces around them.

I learn a lot from that and then I'm a very active public speaker. I give maybe five or six speeches per month and around twelve keynotes per year.

Heather Shapiro

I moved from San Francisco to Boston to be closer to my family and found I needed to build a new network.

My spouse and I had both been in San Francisco for six years and I'd been working in data science and felt pretty well connected in the data science network. But we were going to have a baby and wanted to be closer to family for that phase of our life.

I was lucky because the company I work for has an office in Boston, and so I requested a transfer.

I became involved in the Women in Data Science conference (WiDS), and it has been amazing. It's probably one of the more fulfilling things I've ever done.

It's a global community, and there were conferences in more than 150 countries this year, including in Israel, the Middle East and Africa — places around the world where women in STEM or data are way less present than they are here in the US.

I try to surround myself with people that I can go to for advice and mentorship. One of my main mentors is a male, but he's also a father. And I have some friends in the data field who have become moms.

But I haven't entirely figured out who to go to when I'm struggling with something and need to talk it out. I think most recently I've had my biggest challenge in the past two weeks or so. It's the idea that work seems to get, like this next level of the career is getting objectively harder at the same time that I'm becoming a mom. It's very real and I haven't figured out how to manage that. I'm thinking that the track I'm on now is not what I want while being a mom. It's a lot of emotional energy. It's a lot of hours.

Something I've been trying to understand is would it be possible to be at a company at this level where I could just work a normal 40-hour week? I do inherently really like the type of work I'm doing now. Building out a data science function and leading a team.

But being a functional leader requires a lot of time. And so is it possible to do that elsewhere? Do I have to leave this company to do it or is this level not going to be something that I want as a mom? Because I want to be able to see my son. That's where I am.

Cathy O'Neil

Data science can be incredibly rewarding. You get to think very broadly, you get to learn new techniques, you get to feel like you've built something that works. There's lots of satisfaction in many of those things. It pays well and it's highly respected.

But when you have a family, you have to set limitations immediately when you start a job. For example, if a company is talking about you taking work home, and if they're sending you emails at 10 p.m., you have to make a conscious decision to not respond until you're at work the next day. Otherwise, you'll set a precedent for yourself that you're available at any hour.

Companies have the same behavior as teenagers when it comes to pushing limits. If they can get more work out of you for less money, they absolutely will try and do that because they're in the business of making money.

As an employee, you have to be in the business of taking care of yourself and your family. You have to set those limitations right at the very beginning. If you've got the skills, if you've got the chops, you've got the degree and you've got the experience, then you've got everything on your side.

Mothers can have an impact through data science work

Kate Strachnyi and Kristen Kehrer

"I've always felt that my work was impactful, I may not be saving the world, but insights found through data often lead to material changes. When you're working for a company that values their customers, you are having an impact on the experiences of real people." -- Kristen Kehrer

One of the advantages of the work is that we get to think critically. And women are quite good at this:

"I helped companies figure out where models could help. What is the best use of it (the model)? How do you define the target? That could be so important and make or break the results. Do they have the right data. Is the data biased? I would ask questions that helped the company best use data science in general and then how to implement it. How to make sure it works. How to get it out to the uses and make them comfortable with this black box."– Olivia Parr-Rud

And here, Natalie Evans Harris talks about how thinking critically led her to one of the most influential opportunities:

"How do we improve equitable opportunity for individuals and communities? I went into college thinking about that, and when I left college I was recruited by the National Security Administration (NSA).

"I chose to go to NSA primarily because I wanted the money to go to grad school, and they paid for that. And so then my plan was always to just do that, get my degree, and then go on to do something at a different agency like HUD or somebody much more connected to communities.

"But I ended up staying at NSA because it turned out that there was this whole community in the military that was not using data in the best way possible. I think a lot of that was influenced by the fact that I joined NSA three months before September 11th.

"Then all of these things about the way agencies don't communicate with each other and how we're not using the data and information that we're collecting to drive decision-making just became so obvious and clear. And so that probably was the biggest influence on the direction I've gone in my career for this point."

Messages from the Parents of Data Science Survey

We received some really inspiring anonymous messages from people who completed the Parents of Data Science Survey *and are sharing them here.*

— Kate & Kristen

I am in a privileged position in that I have great control over my time. Most people do not have this level of flexibility.

***

It's been important to clarify my priorities—my kids—and organize my life accordingly. That means I earn less than I would like but am happier. I can make more money later when the kids are at home less.

***

Changed careers from lab-oriented neuroscience researcher to data person in healthcare. Best move ever. I have more time to invest in relationships with my sons while still being intellectually challenged, engaged and thrilled every day!

***

Having bosses that put family first and understand flexibility are key to a happier employee.

***

The company that I am at now goes to great lengths to support family-work balance, including flexibility in when you work, mostly remote work, paying in towards child care benefits, an atmosphere that recognizes and supports a life and priorities outside of work and a culture of respect and empathy.

CHAPTER 2

MOTHERS & DATA SCIENCE: THE CHALLENGES

There is still a gap when it comes to having an equal number of men and women in the data science space. It begins with how we are raised and the environment that surrounds us.

— Kate & Kristen

Underrepresentation and stereotyping

Deborah Berebichez

My interest in science was endearing at 12, but at 16 it was viewed as a waste of time. When I confessed to my parents and teachers in high school that I wanted to study math and physics at university, I was met with disapproval.

I was told those careers were too hard for women, and I should study something more appropriate like communications or marketing.

One way to push back against that negative message is to encourage the next generation of minds to pursue science, especially women and minorities who are underrepresented.

Kristen Kehrer

Bias doesn't come from someone just being "male." It's misogyny that's the problem. If there is no misogyny in the work environment, the sex or gender of your co-workers doesn't really matter.

At my first job out of school they only allowed us to wear jeans on Fridays. And actually in my first review period it wasn't explicitly written down, but when I asked for clarification of one of the items on my review my boss said "We'd like you to dress nicer."

Would they have said that to a man? Probably not. This plays very much into the double standard of how women are expected to act and dress a certain way.

I've also had a boss question my LinkedIn profile picture and ask, *"Do you think this is what a data scientist looks like?"* The answer is of course *yes* because I'm a data scientist, and I look like that.

In school and at a previous job it was mentioned that if someone was male and had the same qualifications as me, I would have been given preference.

So, there may not have been "barriers" to entry, but at the same time, no one was willing to let me forget for one second that I was different from men, being given preferential treatment and that I should be aware that I'm "lucky" or something.

During the interview for my first job, I was not as keen to truly interview the company. I was just trying to land a job. Throughout the rest of my career, though, cultural fit has been paramount to my decision to take a position.

In my last job search I received four offers and went with the job where the team members gave me the "warm fuzzy" feeling. My boss was male, the team was diverse, and the dynamic was incredibly friendly and inviting.

I am also the type of person who just wouldn't put up with poor behavior for an extended period of time. There are plenty of opportunities in data science at progressive companies where people treat women with respect. I promise you that they're out there.

Kate Strachnyi

From my perspective, there haven't really been barriers to entry into the space. I've encountered a few situations where I noticed a lesser representation of women, for example on speaker panels at data conferences. I do think we have a ways to go before we see equal representation in the industry, but have to agree with Claudia that the more we talk about it, the bigger we make it seem, especially with the future generation.

It's really a balance of not closing our eyes and being blind to gender imbalance or harassment and at the same time being conscious of not making the barriers appear more prominent to the next generation.

> "I fear that the more we debate 'barriers', the more we make them appear real to young women, as if making a career in STEM is more of an uphill battle than making a career in general." — Claudia Perlich

My entire career has involved me being one of the few females in the room, from studying finance and taking classes such as econometrics, to working in the financial services industry/banking sector, which is very much male dominated. At times I did feel left out. When all of the team-building events centered around sports (something I just didn't follow at the time), it was hard to keep up.

The discussions around whose team won last night made me feel a bit left out since I couldn't take part in that discussion.

The ability to learn specific aspects of the role, the more technical/detailed aspects, allowed me to prove myself by being the person that had the answers to very specific questions.

Feelings of not belonging or being liked

Perhaps because of indoctrination that they don't belong in science — and societal pressure to be able to "do it all" successfully — mothers in data science can experience the feeling of "imposter syndrome," in which they feel that they don't belong or deserve to be in this profession.

Another uncomfortable feeling was wanting to be liked because of perception or reality that this is the only way to succeed and gain promotions. Some women may even unconsciously place a higher value on being liked than appearing as competent. It's unfortunate because men in society can more easily be perceived as both.

— Kate & Kristen

Natalie Evans Harris

Imposter syndrome is almost like I have to prove that I'm a good mom and a good wife and a good leader. I have to prove it to myself and I also have to prove it in everything I do, and I find myself sometimes overcompensating in order to be able to prove that I can.

Deborah Berebichez

Even though I realized these thoughts of imposter syndrome were not true, I still felt them for a long time. Adjusting to working in the corporate environment was difficult for me. I always felt like maybe I was saying the wrong thing.

But this has changed with time. If speaking up for yourself is hard, and you don't feel like you know how you'll be received, just know that over time these things get easier. It won't feel so stressful forever.

Alice Zhao

When I first joined my current company, my son still wasn't sleeping through the night yet, and I was doing my best to do my job well during the day and take care of my family at night.

My team consisted of a dozen data scientists and they were constantly having discussions about the latest algorithms and tools. While I should have felt fortunate to be a part of a team that was so talented and driven, all I could think to myself was *"I really should be learning this. I wish I knew more. I'm so behind and I don't have the time to catch up."*

Over time, I've learned that I don't have to dedicate all of my free time to learning data science to be a good data scientist. I am confident in the things that I'm good at, which are making complex things easy to understand and mentoring students. I'm just going to focus on those things and not worry about all the other stuff.

One of the mottos we tell our students (at Metis) on day one is if you don't know something, be honest. If you do know something, be humble and teach others. It's ok not to know everything. No one does. Everyone has their own strengths. That should be the focus instead of comparing yourself with others.

Kristen Keher

I had a tremendously powerful experience in 2009 at a Women In Mathematics conference while working on my master's degree in statistics. I was in an auditorium full of women, and the person leading this talk asked, "Do you ever feel like if people knew the real you, you wouldn't be where you are?"

Everyone's hands went up. I couldn't believe it. I thought I was unique. I thought that I felt this way because after finishing my bachelor's degree but before starting my master's, I had made some terrible decisions and as a result my self-esteem was suffering. To see that other women felt this way helped me to realize that these thoughts of imposter syndrome affected most. This helped me to come to the realization that these thoughts of being "not good enough" were not the truth.

It took years to get to a point where I felt confident in my abilities and found my voice.

If anyone is having these thoughts, know that you deserve what you have accomplished. You've worked hard and you've earned your seat at the table.

Being a working mother is stressful

Mothers feel stress no matter what career choice they make. Mothers often feel underprepared and inundated by the demands of having a child (Francis-Connolly, 2002).

Also, women often take on more of the traditional household duties after the birth of their first child (Gjerdingen and Center, 2005), and as household and child-rearing duties increase, many women feel a lack of free time to relax.

Mothers report higher stress, greater depression, and lower self-esteem after the birth of a child, compared to the fathers (Pancer, Pratt, Hunsberger, & Gallant, 2000).[2]

— Kate & Kristen

---

2 http://www.sciedu.ca/journal/index.php/jms/article/viewFile/8668/5227

Deborah Berebichez

Becoming a mother for me was the most magical experience in the world. Yet the beginning was really difficult. I wasn't prepared for the kind of shock to the system that it is, and I think it's partly because I didn't have my child in my 20s. I became a mom when I was older. While that has advantages in that my career was already established and I could afford to take time off and dictate my own terms at work, it was also a very challenging time, because the way I grew up in Mexico City, I didn't really have role models that were working moms. Forget about data science, just working moms.

I felt tremendously guilty the first few months about leaving my child at home in the care of somebody who I didn't know very well. My husband and a couple friends that know me very well said, "Yes, you feel horrible right now, but if you were to quit, in six months or a year, you're going to regret it." So I just kept working.

Kate Strachnyi

In the male-dominated field of finance, it became a bit harder for me once I had kids and no longer had an interest or the free time to join my male colleagues out at night after work. That's when much of the bonding takes place and it was hard to be a part of that since I really just wanted to be home with my kids.

This is why I'm glad I took on an internal role with a team that mainly works remotely and is understanding of me having to be away at times or just flexible with the hours.

Kristen Kehrer

I've certainly experienced personal "mom" barriers — feeling guilty that I need to take time off of work to take a child to the doctors or that my child is sick again. Feeling guilty when daycare is closed, because my job wasn't always around my schedule. I was working for someone else, and they had their own priorities and agenda. These personal mindset barriers I had may not even have been grounded in reality. However, I have found that consulting and being my own boss allows me to more easily set boundaries around when I'm available, making the juggle less stressful.

If anything, I may have benefitted in my career specifically because I am a woman. And I'm OK with that, too.

CHAPTER 3

CARLA GENTRY: IN HER OWN VOICE

Carla raised two sons as a single mother as she attended college for six years and then started working as an analyst and data scientist in the late '90s. Today she is an advocate and educator and helps companies use data science to create strategies for better customer experiences. This is her story of being a mother in data science.

— Kate & Kristen

When I got out of college, my kids were 10 and 13, impressionable ages. We lived in a little suburb right outside Chicago. They had access to the best schools, some of the best schools in Illinois at the time. They were getting a good education, so I thought I was doing my part as a parent, but I was ignoring the little things like the conversations that you have at dinner, asking them how their day was.

I think it's hard as a woman. We're pulled in two directions. We're innately a mother. We always mother our kids, our friends. It's the way we were born, to take care of others.

But then, we also want to be successful. We also want to receive those accolades for the things that we have accomplished, and I think that, too, is frustrating.

You do sacrifice one for the other. I think by the time I realized that I had missed a big chunk of the kids' teenage years, my eldest one was joining the Navy and moving out. And my youngest one was 17 years old and didn't want anything to do with mom. He wanted to be with his friends.

I do have regrets. I wish that I had not been a single mother because that's the hard part, when it's just you.

Basically I quit high school two weeks into the 11th grade. I got married when I was 17 and had my first son when I was 19. I had my second child when I was 22, and then I got divorced six months after he was born. I had a newborn baby and a 3-year-old, and I was 22 years old.

You do what you have to do to survive. I think a lot of mothers out there will admit this to themselves, that you can't think about what you're doing, because if you think about it, you'll fail. You just have to do it.

People would come up and pat me on the back when I was getting ready to graduate and say, *"Oh my god, I can't believe it. You've been in college for six years, you're going to graduate with a degree in math and a degree in economics. How did you do it?"* And I would say I have no idea. I just refused to quit.

Why we need women in data science

Right now data science is a male-dominated field. The male opinion can often overshadow a more diverse view. And as we know with machine learning, we're always working with historical data. If there is bias in history, these models will be perpetuating biases. It may take 50 years for us to have more evenly mixed teams, but if we have a team that's 70/30 or 60/40, or whatever, we get that contribution from women as well.

Diversity of thought on machine learning teams will help ensure that the voices of women are heard, as they do represent such a large part of the total population, after all.

The diversity of the team will make the team better because we know that men and women don't always think alike. Are we logical? Are we creative? You get that, when you have a team that's men and women, you get such a great mix of intelligence. You get that macro intelligence that I love, because you're getting other people's opinions.

Also, most of the time when you think of logic, people will think of men. I don't know if that's necessarily true, but it is a bias that is perpetuated through the media. Women are quite logical, because in a crisis situation with a child, my husband always was over in a corner curled up in a ball, where I was the one handling the situation. Although this is not always the case.

I also think women are natural problem solvers, because we face those opportunities for extreme multitasking every day. When you've got kids, you're trying to juggle: *OK, this one needs lunch, this one needs nursing, this one needs a diaper change, this one is screaming at the top of his lungs.*

Although it is not true in every home, statistically it is most often the woman who is expected to be calm, cool, and collected. Women are natural problem solvers.

Search until you find the right job

Data science is a great field, but you have to go into it with open eyes. When you have a family, you have to set healthy boundaries immediately when you start a job.

You've got little ones, and your boss is just going to have to understand your kid is sick, and you're going to be working from home today, period. This may seem like an obvious necessity, but women are more likely to feel stress around expressing their needs when it comes to taking time to attend to a child's needs.

"This difference was evident in our survey. So although there are fathers who feel this pressure, it's clear that it's more often experienced by the mothers". - **Kristen Kehrer**

You need to interview the hiring managers and find out if this is the company where you want to spend your life. What kind of perks are they going to give you? What kind of love are you going to have at this company?

We spend a tremendous amount of our awake time that is away from our children at work. Do the bosses and coworkers seem like people who will understand that you have to leave a meeting if you get a call from school that a child is sick? Will they understand if your child is up all night throwing up that you're going to work from home today? And will you be able to work from home without feeling like people are judging you?

And if that job doesn't want you, then that's OK. Move on to the next job.

It's easier to find a good fit the first time than accept a position that is showing obvious red flags and have to find a new position. No thanks! Listen to your intuition on whether or not you feel like the people you'll be working with will accept you, and are willing to provide a work-life balance that aligns with your life.

Part of this is that we need to advocate for our needs because no one can read our minds, but you typically have a gut feeling about whether someone is willing to work with you. Often we figure that out by making a simple request regarding our needs and noticing the response.

You also have to become a subject matter expert. If you're going to talk about something, you have to talk about it with expertise. You can't just write a short, little snippet blog and expect everybody to receive it like it's golden.

For me it started when I went down to the IT department and begged one of the IT guys to pull my own data. So you have to find a liaison, whether it's male or female, to help you. Find somebody who will teach you what you don't know.

How things have changed

If I look back 21 years when I first got started, I had people who wouldn't even talk to me. At least now I get some respect. I don't get as much respect as the guys do, but at least I get more respect than I did back then.

Early on I felt like no matter how hard I worked, it was my boss, who's a great guy, who was given all the credit for anything that I'd accomplished because he was my boss. And I saw that throughout my career. People got credit because they were the VP, or they were the managing director, or whatever. It always seemed like the women were in the shadow of a guy.

Right now it is, for us, an employee's market because there's just not enough talent out there. There's not enough STEM graduates, there's not enough people with logic and mathematics backgrounds. These are what's important.

Computer scientists are fantastic, don't get me wrong. I love them, but they do not have business sense. And they'll tell you right up front, *"Hey, I'll fold the data, but what you do with it after that is your business."* They're not in the business of analyzing data.

But data scientists and advanced analytics people, statisticians, that's our jobs, that's what we do.

What success means to me

Success for me is recognition for being a subject matter expert in my field, to have 385,000 people following me on LinkedIn and saying that I made a difference in their lives. Also, to know that I am making a contribution to the data science field and that my comments are received and welcomed and that I'm able to help people.

We have to take time for ourselves and pass this information of the mistakes that we've made and the things that we've learned off to the next generation so that they don't make those same mistakes.

It's about hope. It's about showing that there is a successful woman out there that's been doing it for 21 years. We do have the power to get out and change these things.

Here's what we need to start doing. If you interviewed with a job and you felt that you deserved a second interview, write to their HR department and make them tell you why you didn't get that second interview. If they still don't want to talk to you, contact the Equal Employment Opportunity Commission.

Something crappy shouldn't have to happen for us to stand up. We should be standing up for ourselves every day in talking about the benefits of having women in the workforce and how it adds an extra dimension to thinking.

But we also have to realize that our life is not infinite. We're only here for a short amount of time. When you're sitting back and you're looking at the photographs of the things that happened and you can barely remember them, you have to think back, *"Well, what is really important to me? Is it more important to me to put my job first? Or to put my family first?"*

Throughout the 21 years, I've come to realize that jobs come and go.

CHAPTER 4

MOTHER/CHILD EXPERIENCES AND STORIES

Pregnancy and birth can be a physical, emotional and mental stressor in many ways, and not all jobs and workplaces are accommodating, despite the high number of women in the workplace. Our experiences and the experiences of the women we interviewed revealed that data science jobs are hit and miss in this area.

– Kate & Kristen

Pregnancy and Birth

Kate Strachnyi

When I was pregnant with my first child, I was not traveling much anymore, mainly working from the office and large financial services institutions (surrounded by men).

I remember sitting in a large conference room at about eight months into pregnancy, and the baby was moving *a lot*. You could see the belly really moving. People stared at me and asked if I was about to have the baby…and every day after that I'd come in and hear the same comment, *"Did you have the baby yet?"* Even when it was clear I had NOT had the baby.

Some of the more difficult moments of pregnancy were during the travel to and from the office. I took a bus and New York City subway train. Most of the time the different smells and standing-room-only space made me a bit nauseous. It was a difficult experience.

With my second child, I was already working remotely and didn't have the same challenges. My challenge was I had an adorable one-and-a-half-year-old running around the house.

Kristen Kehrer

I had difficulty getting pregnant with my first child. Roughly 11% of women have experienced difficulties trying to conceive (National Institute of Health https://www.nichd.nih.gov/health/topics/infertility/conditioninfo/common). That's a whole lot of people who are trying to deal with something quite emotional, so I'm typically open with my experience, hoping that somewhere out there is another couple that has the opportunity to feel less alone as a result of my sharing.

When it came time to see the reproductive endocrinologist about fertility treatments, it definitely added stress to an already emotional experience.

With fertility treatments, you need to go see the doctor on very specific days in your cycle to have blood taken, plus visits for treatments. I was trying to manage this all while on synthetic hormones without telling my boss that this is what I was doing.

Trying to make it to appointments before work started, trying to get to appointments that might make me physically uncomfortable for the afternoon and not being sure whether or not I'd make it back to the office that day — it was all added stress that most likely wouldn't have existed had I just shared what I was going through.

My boss probably wouldn't have even batted an eye. He had a son, we were in the healthcare industry, and although he probably wouldn't want to have a heart-to-heart about it, he'd be fine knowing the reason why I had appointments.

Just before I became pregnant, I had started a new job as a lead analyst. Luckily it was OK that I had been working there for only two months. I still qualified for paid maternity leave through short-term disability, and it was paid at 100% for 10 weeks.

I was so excited to be pregnant that nothing could get me down. Not the morning sickness, nothing. Then at 20 weeks I was diagnosed with gestational diabetes and was on insulin three times per day.

This would mean additional monitoring at the hospital, extra ultra-sounds, and again, I was dealing with the stress of being pretty new to a role, needing extra time off, and feeling plenty of guilt and stress.

After a rather traumatic birth experience, I had a wonderfully healthy baby girl. Then came the post-partum depression (PPD) and anxiety. Again, this is a taboo topic that is not often talked about openly.

It took me a while before I realized I needed help. They often talk about the "baby blues" after giving birth, so it's hard to tell when you've reached the point that we're no longer talking about baby blues.

I didn't have any previous therapists, and every place I called was quoting me a minimum of weeks before I could get in to see a psychiatrist. All the while, I also wasn't able to care for my baby the way that I wanted to be able to. I finally just drove to the ER with my baby and decided I wasn't leaving until I got to speak with the psychiatrist in the ER. This was immensely helpful in getting the ball rolling with PPD meds that worked for me.

Unlike many moms who are sad to return to work after a baby, I was so ready. It was definitely because my PPD was not being fully managed, but I felt a tremendous amount of guilt that I was happy to return to work.

The director of analytics took me aside on my first day back and said, "I know that coming back to work after a baby is difficult. If you need to leave early, or talk, or if you need anything, just ask." I looked at her and said, "I'm happy to be back," and continued with my work. This also caused me a ton of guilt.

Luckily, we knew the deal with my second child. As soon as I had my second baby, I immediately started on PPD meds. It gave me the opportunity to truly enjoy my maternity leave. And then again I felt guilty that I was able to enjoy my second child's maternity leave more than my first.

So I guess the theme of my pregnancies and birth is just guilt, guilt, guilt. But it didn't last forever, and I have two amazing children that I love so much as a result. I feel so fortunate.

Claudia Perlich

I grew up in East Germany, but I just got lucky that the timing of the wall coming down was perfect for me to go off and study in the West. Ultimately, I found my way to New York in '98 to start my Ph.D. at New York University.

I had a very good Ph.D. experience. I could have finished after five years, but I decided to have my son in the sixth year. I was done with a lot of the kind of technical work by the time I even got pregnant, so mostly at this point it was about writing up my dissertation and doing a few more experiments. I had an extremely supportive adviser. He was just a little bit older and had children.

And then being on the academic job market, which means you need to go to the job market conferences. My son was born in March, and the conferences are in December, meaning I was six-and-a-half-months pregnant. But I was feeling good.

After that you have all these kinds of trips. You're flying out for interviews and I did five, six, seven of those. Iowa, Boston, Maryland. I did those and that was still pretty much OK. My last interview was literally two weeks before he was born. It was a two-day interview at IBM Research and that's the job I took.

I was the elephant in the room because at that point, it was pretty damn obvious. I was interviewing with the Predicting Modeling Group. At the time, the woman who was heading that department had about 100 people reporting to her. She had raised her kids while building her career at IBM, and when I gave birth to my son, I remember getting a hand-written card from her congratulating me. And that was even before I had ever accepted the job, and I really felt like, wow, they get it.

It seemed a lot less competitive. You didn't have these debates about whether you can extend your tenure track by a year if you have a kid. None of this was really a question there. People just lived their lives and they all have very decent work-life balance."

Maternity Leave

Maternity leave policies vary based on geography and specific companies. Having children in general in the U.S. isn't something that appears to be planned for by companies. Women can feel pressure to stay in their current job to take advantage of paid maternity leave.

An interesting but honestly not-surprising finding from the Parents of Data Science survey shows that 73% of women in the U.S. wish that their maternity leave was longer. These early weeks/months of bonding time with your baby are just so precious. It's sad that it is considered a "privilege" to do the one thing that should be the most natural – spending time with your newborn.

— Kate & Kristen

Natalie Evans Harris

There was definitely not a point in my career in the intelligence community where I would have been comfortable taking so much time off work. At that time there was no way I could stop what I was doing and take three months off or take six weeks off. It just would have been unheard of and it definitely would have disrupted my career path. I saw it happen to other women.

Things are different today. Women in data science are fortunate and should capitalize on the fact that they can work remotely. If having that remote flexibility in your work is important, there are enough opportunities out there.

This flexibility is not a corporate environment accommodating us. It's just the way it should be because healthy families lead to healthy societies.

When I co-founded Right Hive, one of the things we as owners discussed was that it was going to be a remote-first company.

Everybody works from home. Our culture is, look, you're a grownup. As long as you get your job done and you are where you're supposed to be, then we're not going to ask you to submit time sheets or to punch a clock.

Kristen Kehrer

I was laid off a week-and-a-half after returning from my second maternity leave. It was truly a blessing, as for the first time since graduating from grad school, I had the opportunity to evaluate all the options available. Data science is such a high-demand field that my schedule was full of interviews in no time.

Looking back, it's definitely funny. When I was pregnant both times I felt "stuck" in my position due to having paid maternity leave. As though I wouldn't be able to find another job with maternity leave while pregnant. Ironically, my boss and I had started working for the company around the same time and my boss had taken the position while pregnant. She obviously felt comfortable making the jump, but I did not. Looking back, for myself, I believe these fears were unfounded. Again, because the demand is so strong for people with data science skills who can communicate well with the business.

Kate Strachnyi

When my husband and I decided to have kids, we had to make sure that I was at a company that provided maternity leave. I was lucky enough to work for a company that provided four months' paid leave. (It's now six months!)

It really hurts me to hear that it's not unusual for a new mother to leave her baby with a nanny or at a daycare at around 6 weeks old to get back to work for fear of losing her job or not making ends meet. There's nothing wrong with a woman doing that, but I think it should be her choice, not by desperation.

The Power of Remote Working

I've been mainly remote for over five years now and feel lucky every single day to have such flexibility. The time saved on my commute alone is around three hours a day — that's 15 hours a WEEK! Just think what you can do with an extra 15 hours each week. I also have the luxury of seeing my daughters off to school and being there for breakfast without rushing.

– Kate

Jacqueline Nolis

I didn't worry for a second about work. As the non-birthing parent I took only four weeks off, which I think, in retrospect, I should have taken more. But I was not worried about work because the fact of my position was, as a director, it was really my job to make sure the team would run effectively when I wasn't there, and so part of it was like, by being gone, this is a test.

I felt confident that they would be able to handle it without me, and I really wanted to focus on my son during that leave, because I know I'll never get a time like this again and it's kind of amazing.

There's like so many little magical moments in that, it felt foolish for me to worry about work. I took my work off my phone. I deleted any way of figuring out what was going on at work so I did not, for a second, think about work.

Cathy O'Neil

My first two kids I had when I was at MIT in a post-doc position that was named for a famous instructor. It was the C.L.E. Moore instructor position, and none of the previous position holders had ever had a child, even though it was 50 years old almost by the time I got there.

I was lucky enough, I guess you could say, to negotiate the maternity policy when I got there and I was already pregnant. So I got a pretty good deal, although it sucked in the sense that there's no forgiveness in academics for a loss of time. The academic schedule is ridiculous.

But I still got a good job at Barnard. But then I went into finance. I was a quantitative analyst at this hedge fund, and they had a pretty good policy for my third kid, which was like 12 weeks of paid leave.

When I became a data scientist and I went to a startup, there was no maternity policy. And I started saying, *"What would it be if I got pregnant?"* I didn't have any plans to get pregnant, but I asked for the sake of all the young women around me. They coughed up something really shitty, like eight weeks part paid or something. It was so weak.

All these young women, and they're the ones that have the least amount of stake in the company, they're getting paid the least, and they have shitty maternity leave. And I think that's typical for smaller companies and nonprofits.

Returning to Work

Heather Shapiro

There was so much buildup to that first week going back to work for me because of not knowing how it was going to be and some anxiety around that.

For my first week back we had my mom come and live with us that week. In preparation I laid out every single outfit for the week, Monday through Friday. And all my bottles were in a row and washed. It was like the first day of school because everything was set up and ready.

It went great. And I felt I was rocking it. But the second week I didn't do any of those things, and then we didn't have my mom.

I was trying to figure out what to wear. Suddenly the bottles were dirty. And so the second week back was so hard, and it was very humbling.

Kate Strachnyi

The first few weeks back to work were truly humbling. You're trying to look put together and on top of everything while simultaneously thinking about what you might be missing at home. Trying to be a good mom, wife and employee all at once.

After returning to work with my second child, I still remember sitting in on a call trying to catch up with all that I missed and hearing that the direction we are going in had changed and the manager that I was reporting to before my leave was no longer there.

Everything had changed and to me it felt really abrupt because I had been away for a few months. It's all about being flexible and adapting to the changing environment.

Pumping/Breastfeeding

This topic doesn't get discussed often, but we feel that it's important for us to share a few stories because it's part of the process and may at times present added challenges to the lives of mothers.

— Kate & Kristen

Kate Strachnyi

With my two kids, I ended up breastfeeding for about a year each and had a different experience each time.

This process is difficult enough on its own without the added complexity of coming into the office and looking for a private place to pump. The first time I returned to the office — about five months after having my child — and I brought in the heavy pump, along with the bottles and various attachments. I also had to carry my laptop back and forth from the office, so it was not a light pack for me that day.

I learned that the office had a "mothers room," which is code for a small room with a chair and table. The room had to be booked in advance like any other conference room at the office.

It was a bit awkward at first because I had to tell the young man at the reception desk that I needed a room to extract milk from my body. I booked the room and went about my business. Another time I recall being in an airport and not really finding a room, so I had to pump in the restroom. It was pretty gross in there.

My ultimate most-memorable pumping experience was when I was working from home and was on a conference call. The call was with seven or eight others, all males, and we were talking about the project status update or something like that and I really had to pump. So I figured, I'd go on mute, and go about my business. Then I'd stop the machine when I had to provide my update and it'll be easy.

What *really* happened was I started the process and forgot to hit the mute button. The machine was pretty quiet, but you can still hear rhythmic sounds. So one of the men asked what the noise was, and everyone on the call started asking. I stayed quiet and finally went on mute.

I guess they eventually realized I'd gone on mute and put two and two together, or they just didn't care anymore and moved on. I felt pretty uncomfortable but quickly decided that I don't have to explain myself. If you're finding yourself in a situation where you might not be comfortable, just remember that this is the most natural thing you can do as a mother, and this will pass quicker than you realize.

At this stage where my kids are 3 and 5 years old, I barely have memories of how difficult the first few years of motherhood had been.

Kristen Kehrer

I was so excited the first time I successfully breastfed in public. We literally texted my sisters-in-law to let them know. It was a big deal. After two months of struggling, I was finally able to do something that I thought was going to be pretty natural. And luckily, this was one week before we had a Disney World trip planned.

With my first, although there was a mothering room at work, I preferred to pump in my car. This was my time to jam out to tunes and escape the office for 15 minutes.

Most health insurance plans in the U.S. give you a breast pump, so I had my healthcare-provided breast pump and then I bought a second breast pump to make my life easier. It was easier for me because this way, I could keep a pump in my car and one at home, so that I didn't need to cart a single pump back and forth to work and then bring it in the house at the end of the day.

I also used Freemie Cups for when I was driving. This was an additional purchase, but so worth it. I would get in the car to head to work, my husband would handle daycare dropoff, and I would pump on my drive to work, even in the drive-through window line to get my breakfast.

They were discreet and a huge time saver. Then I'd keep a cooler in my car with ice packs to store the milk and pump parts. I think the official recommendation now might be to wash pump parts after each use, but at the time, the refrigerator or kept cold was considered sufficient between pumping sessions.

Feelings of Guilt

According to the Parents of Data Science Survey, 52% of women reported that their spouse handles at least half of the home/baby care responsibilities. This was an inspiring data point to uncover!

Kristen Kehrer

My husband handles at least 50% of the actual childcare responsibilities. He is also an equal partner in keeping the household afloat. However, I do feel guilty whenever I travel or anything else that doesn't allow me to be home in time to prepare dinner on any given evening, because I'm not there to assist with sharing the load.

Kate Strachnyi

I feel mama's guilt from a few angles. My mother and I share a home, and she helps out with a large share of the housework and child care.

I feel guilty that I don't help out more around the house. I feel guilty when I don't see my kids off to school or help them with homework once they are back. I feel guilty when I take time off work to go take my kids to the park because I feel like I should be working.

At the end of the day, the guilt sort of equals out and I understand that we are human and have demands on our time and schedules so I forgive myself, smile and start the process all over again the next day.

Quotes from the Parents of Data Science Survey

I was laid off shortly after returning from maternity leave and had to deal with difficulties helping leadership establish a proper mothers room in the office. It resulted in disrespectful treatment and unnecessary stress.

*****

I only just returned to work after maternity leave (no maternity leave with work; only state disability). Right as I returned, my child got sick from daycare. My company allowed me to work from home while he was sick and will continue to do so, but I still felt guilty asking.

*****

Having management that also has kids makes such an incredible difference. My previous supervisor had no children and no concept of what a newborn required.

***

Though startups rightfully offer more opportunities to work on new-age technologies like data science and cloud computing, they have a horrible work-life balance. Not only do they expect you to take your work home, but somehow we need to share the same passion for their company which they have.

CHAPTER 5

WORDS OF WISDOM

Kate Strachnyi: Reach Out To Find Your Place

I truly believe that working with data is a great option for mothers. The self-satisfaction I receive from completing a project and knowing my work has made a difference means a lot to me. Additionally, the role is typically fairly compensated and you get to continue learning and growing (which is a big motivator for me). If this is a space you enjoy working in, I say jump in with both feet and go for it.

One thing that has really worked for me was reaching out to people on social media (mainly LinkedIn) to receive advice on how to get started, or what to learn, among other advice. I've actually ended up writing two books that summarize the advice received from the kind individuals that have volunteered their time and effort to support this initiative: *Journey to Data Scientist* and *The Disruptors: Data Science Leaders*.

I also began posting on social media late in 2017. I would learn a new concept in data science and share an article or write a quick post about the topic to help the community also learn this topic, which later transitioned into putting out videos and developing a YouTube channel (Story By Data).

This effort really helped me understand and grasp the concepts that I was trying to learn, and also helped me grow my network in the data community. Eventually I was interviewed for several data science podcasts to try to help spread the knowledge and experience.

One key thing that came out of the social media posts (in addition to several consulting gig offers, speaking opportunities, etc.) was the ability to build real friendships and relationships. I actually met Kristen Kehrer on LinkedIn, and we've gone on to write this book together.

My advice is to use the resources that we have at our fingertips to not only consume content, but to also produce content and tell our story.

Cathy O'Neil: Be Realistic About Expectations

I'm grappling with the fact that as a field, I'm not sure we're actually making the world a better place. I welcome women who are thoughtful and will improve the field, but I also realize that women walking into this field will probably not be empowered to decide where it goes. So they're going to be told what to do, not asked what they should be doing.

I have no regrets personally, but that doesn't mean everything I saw was peachy keen. Let me put it this way. When people ask me, *"Hey, I wanna be a data scientist. Where should I work to make the world a better place?"* I don't have an answer for them. And I really wish I did.

Instead, the best I can do is, "Well, go to Google and Facebook, learn what they're doing, be a whistleblower or use the techniques in the future to do better. Because they're pretending to make the world a better place, but there's actually a profit motive that is much, much, much more real than any motive of fairness or ethics.

That's not to say that everything that's going on is unethical. That's not what I mean. It's not. It's perfectly fine and mostly benign and mostly annoying at worst. But it's not explicitly trying to do good. Data science, generally speaking, makes rich people richer and poor people poorer.

Mixed Feelings: Work-Life Balance and 'Having It All'

We've always loved hearing the stories of women in tech. I love hearing about different career paths, overcoming obstacles and about being a savvy badass. While there are many stories of women in tech, and there are many initiatives popping up aimed at highlighting women's contribution to tech, you hear much less about what it truly looks like to balance being a parent and a tech career.

— Kate & Kristen

Natalie Evans Harris

My husband said, *"How would you feel about being a stay-at-home mom?"* I was like, *"Well how would you feel if I died?"*

Heather Shapiro

I don't think it's a myth. I've had work-life balance before at prior companies. I think I hit the sweet spot working three days a week when I was a contractor. I didn't have kids then, but I know that is a possibility.

I don't have the work-life balance I desire right now. I think I can get back there, but it's going to take some work and trying to sort through how to do that while still feeling like I'm otherwise moving forward in my career.

That's the question I keep asking myself is do I care about moving forward and what does that mean? I think I don't care enough about my career. In the scope of life it is not as important to me as other things.

There is one woman at my current company who is a VP whom I've identified as a mentor. I have been talking to her about this and the challenges I'm having.

She's been an amazing support person and has been trying to help me. She doesn't want me to burn out and leave. She's working with me to see if we can figure something out with where I am now.

Alice Zhao

I think it's possible to have it all, but in moderation. I'm not at the top of the corporate ladder, but I enjoy my current job and my company treats me well. I'm not a Pinterest perfect mom, but I get to spend time each weekend playing and baking with my kids.

There is less "me" time, less time for socializing, date nights, exercising and sleep. But it's all about the moderation, right? It's the reality of a working parent and I've come to accept it. I do feel fortunate that I have a very supportive husband, which is more important than I ever thought it would be.

Jacqueline Nolis

I am working on a huge consulting project. I have a second consulting project on the side. I'm writing a book. (*Build Your Career in Data Science*), co-authored with Emily Robinson. It's about how to get into data science careers.

And I'm being a parent, so I'm doing many, many things at once, and it is so difficult.

I've actually gotten into this groove where I'm making it work, but then I look at some of my friends who aren't parents and just have a standard 40-hour-a-week job and don't have anything on the side like that, just every weekend they get like 48 hours to just do whatever they want. I die a little inside because I think, that used to be me. I could do a lot. So that's really hard.

I'm going to finish writing this book. I'm going to continue to raise my son and go to work, and I'm going to make it happen, but it's just that you lose this really important thing of just having a certain amount of freedom.

Other Voices on Work-Life Balance

The Media's Impact

"The choice of whether a mother should work if she is financially stable is one more issue that causes debate and brings forth the concept of a good mother versus a bad mother. The media show the progression of the stay-at-home mother of the 1950s to the current version of the mother who can do it all. This do-it-all mentality entails managing a successful career while maintaining an idealistic portrayal of a mother."[1]

Evidence of 'Role Strain' Is Not Conclusive

"Role strain" refers to the demands of taking on multiple roles and the coping strategies used to deal with the strain (Goode, 1960). Researchers differ in how much role strain mothers have. Many women feel role strain when they combine work and family (Aneshensel & Pearlin, 1987). Other women have felt the dual roles of both mother and employee outweigh the strain and provide them self-worth (Marks, 1977).

Kristen Kehrer: Find Help Where You Can

I think we don't talk enough about the help we have. I feel I have it all, but I'm also not able to do it all. I have a housecleaner, at times we've paid people to prepare our food, and we have our groceries delivered.

We go out as a family to eat probably twice a week to make life a little easier. We sometimes have a dog walker to make sure that our beloved corgi receives enough attention. And finally, both my husband and I have parents that live about 40 minutes away, so although they're not watching the kids every day, we are able to get a date night on occasion.

Our daycare center also is incredibly reliable. I didn't have as much help in the beginning, but as you work and make more money, those extra expenses that make life a little easier become very worth it. I'm also very fortunate to be in a dual-income household. My husband is also paid well, and this helps to make these niceties possible. If I didn't have help, I think I'd be much more tired, and I don't believe I'd have the time to be writing this book.

Kate Strachnyi: Look For Family Options

I had a conversation with my husband about a month after having our first child about the options that we have for childcare, work, etc., and we made a list of options together including – hire a nanny and work in the office, put her in daycare and work in the office, work from home and try to also handle the baby.

Then he suggested that we hire someone to help me at home while I try to find a work-from-home gig. Which is exactly what we ended up doing! My mom was working as a nanny for another family and I asked if I could just hire her to stay home, and she said yes. It's been a perfect match ever since. I never have to worry about trust or privacy of someone else being in my house.

Jacqueline Nolis: Wherever You Are, Keep Learning

I am done with school. I've gotten as many degrees as I need, so I haven't been doing formal education, but part of the data science job is constantly learning new skills.

The best way for me that I've found to learn new skills is just really conveniently to try and pick new things up either on the job or like on a side project. And so I continue to learn by just constantly being at my job being like, oh, wow, there's a new skill I need. I'm going to go learn that while on the job or while doing something fun on the side. It's just making sure I've had space to either pick those things up at work or, if needed, just try to do something fun on the side to learn something new.

There are lots of places in data science where you learn the best from doing things and experimenting and trying things on your own, whether on your job or on the side.

Data science is unique in that there are a lot of people telling you that, that's not the right way to do it. People say you need a Ph.D to get an AI job. You need to do a boot camp. Well, how can you do this without having taken eight courses in linear algebra. There's a lot of people saying you need this formal education and so I think data scientists are especially vulnerable to being like well, I'm not ready to try this out because I haven't taken enough courses yet.

Perspectives on Fatherhood

Kristen Kehrer

Mothers/future mothers may worry that having children will make it difficult for them to fulfill their career aspirations. It's worth noting that dads/future dads don't really need to consider if it's possible to have a great career and be a great dad. This isn't in every case but is worth mentioning.

And although some men do think about the challenges of balancing a family and their career aspirations, they do not typically need to take time off for all of the doctor's visits during maternity leave. They also don't have to worry about leaving a meeting to pump, scheduling often scarce mothering room resources or having breast milk leak through their shirt.

Jacqueline Nolis

On the one hand, I don't have an accurate description of this or knowledge of this because I'm in a lesbian relationship, and so there are no dads, but I used to be a dad.

My wife did a really great thing while she was pregnant. It was about four or five months in, and she called me out. She was like, Jacqueline, look, and I was still a guy at the time, still male presenting; she was like, Jacqueline, look, I'm doing all this work and I'm planning all the items we're going to buy for, you know, and I'm doing all this planning to prepare for when we have a kid.

There's nothing that, right now, you are the captain of. That's not equal, and we can't have a scenario where, once Amber is born, it continues to be unequal either, so it needs to stop right now.

And I'm like, oh, shit. And so, that kind of got me started on OK, I'm going to captain finding a preschool. I'm going to captain finding the au pair. It started me captaining a little bit and getting into the mindset of like, no I need to take responsibility for this because this is part mine.

This really relates to being trans because I think a lot during this process was me deciding that I don't want to be a guy. I don't want to fill that role. I want to have a lot of things to do, so by the time Amber was

6 months old, I had transitioned. The relationship got quite equitable so things are very even and so I feel very happy now with how we distributed the labor.

47

CHAPTER 6

DEEP DIVE ON SURVEY

About the Survey

The data science community was so generous to share this survey. The first post of the survey on LinkedIn resulted in 32 reshares (survey was first posted in May 2019). We continued to share the survey for two weeks across social media. Once we had over 300 responses, we decided to start looking at the data.

Caveats

Before we talk about the specific questions asked and share some of the results, we need to have an important PSA about bias:

This survey was obviously not a random sample of parents in data science. We shared it on our social media, and we're both established in our careers. The audience that was most likely to see the survey might have been skewed toward more mid-career folks. Also, there is probably a whole world of data scientists who do not spend their time hanging out in data science social circles on social media. Plus, response bias. No idea what type of person is more likely to take or not take this survey for any other random reason.

If you'd like to add your personal data points as a parent in data science, the survey is still live and a link can be found on the resources page of this book.

Even though the data is biased, it's still interesting to look at. We just want to be clear that we're not making any claims about this data being representative of the population of parent data scientists.

Results

With the caveat that this survey is most likely quite biased, there were still a number of results that we were so pleased to see.

1. 47% of the respondents identified as female. Since we know there is a gender disparity in tech, this was a pleasant surprise. However, we have both been cultivating relationships with women in data science, so our network may have a higher proportion of women than the actual data science population.

2. 71% of men want longer parental leave. This number becomes 74% when we look at men in the U.S. We love that men want to be home during this time.

    73% of women in the U.S. wish their maternity leave was longer. We all know that the current state of paternal leave in the U.S. is quite sad. We were hoping that being in a field that is known for pretty decent working standards would have happier employees regarding their paternal leave. Unfortunately, this is not the case.

    We do see that the number of men having two weeks or less of paternity leave is quite high, so it's understandable that more time is required to fully acclimate to the life-changing event of having a child.

    3. 52% of women reported that their spouse handles at least half of the home/baby-care responsibilities. I wonder if data science is a more progressive field than others, if this is simply due to bias in our following, or if the world as a whole is becoming more progressive. Regardless, it's an interesting result.

Women were much more likely to report being treated unfairly at work compared to men:

The age distribution was more widespread for men compared to women. There were more men with a longer tenure in the field, and hopefully that is what accounts for the men making more money.

Due to the sample size of 332, it is not possible to drill down too deeply into the data.

You can see the dashboard for yourself at https://datamovesme.com/data4datanerds. Full disclosure: Kristen built this dashboard in only two days. Many improvements could be made in terms of the presentation of the data, but we believe this works for our purposes.

[1] http://www.sciedu.ca/journal/index.php/jms/article/viewFile/8668/5227

[2] https://www.mckinsey.com/business-functions/organization/our-insights/facebook-sheryl-sandberg-no-one-can-have-it-all

CONCLUSION

HOPES FOR OUR CHILDREN

Jacqueline Nolis

My biggest hope for my son Amber is that he's really happy with himself and his life, like he's really happy with whatever he does. Growing up, I had a lot of pressure on me to be successful and to get married with a house and the picket fence. As an adult I have been so focused on success that I don't really know how to be happy.

And so, the thing I really hope for Amber is that he really just is able to be in a place where whatever he does, whether it's career success, family success . . he's happy with it.

I guess the sub-goal and for us related to that is that he is really just able to have good relationships with other people, that he has friends, that he has strong bonds in his life, because I think that's a lot of what drives happiness.

Kristen Kehrer

I am planning to expose my children to coding early. It may just be that both myself and my husband are always coding. I'm always wildly excited to explain how coding works when my daughter asks what I'm working on. I certainly wouldn't want my children to think that coding is out of reach for them. I also want to give them room to find a purpose that lights them up (even if it doesn't involve STEM), but I will be a voice that tries to challenge what they'll hear in society. Whether or not "Math isn't fun" is someone's experience should be a decision everyone makes on their own, not just a message they're constantly bombarded with in society. I'm hoping my daughter is confused and befuddled the first time she hears the stereotype that "girls aren't good at math".

Deborah Berebichez

It's such a complex time for women, and for men. We're trying to re-educate a generation on how to be truly caring citizens, and what we consider the basic requirements of an ethical project. We are struggling with similar questions in data science, how our algorithms can do good in the world while being fair and without invading privacy.

What I hope for my daughter is that she grows up to be a critical thinker, that she has the freedom to choose her profession, her lifestyle, and that she learns that ethics matters a great deal, that to be a good person in this world, and to have a big heart are way more important than fame or any money she'll ever make. I want the same things for my son of course.

Claudia Perlich

I think it's really important that you are true to yourself and that you figure out what you're really good at. My philosophy has always been to keep doors open and make sure you have different options to choose from.

I'm not saying that everybody has to be a data scientist. Is that something that you enjoy and that you consider doing? Absolutely go for it, because I think there's a huge market moving forward, and it's not going to go away. I don't see that disappearing in the next five to 10 years.

But you know what? If my son decides he'd rather be a gardener, that's his choice to make, and I'll try to support him in that one too.

Nobody has the same job for 30 years anymore. So having things that are valued in a large kind of breadth of different industries and so on, I think that's a very good bet moving forward.

Cathy O'Neil

I have conflicting hopes for my kids. I want them to be fulfilled, but I don't want them to be assholes. But I also don't want them to be defensive, but I want them to be careful. I look around at the #MeToo stuff, and I'm like this is really, really, really important, and yet I don't want my children, my white boy children to feel like they've already fucked up.

And it's this weird, it's a weird time. I also don't want them to feel entitled, but I also want them to succeed. It's a really tricky message. I want them to be really good at something, but I also want them to figure out what they actually care about. I don't know. I guess my overall goals for my boys are somehow the same goals I have for myself. The jury's out.

BIOGRAPHIES OF PEOPLE WE INTERVIEWED

We suggest you follow these awesome mothers on social media to stay up to date on their journeys.

— Kate & Kristen

Alice Zhao

I am a data scientist who is passionate about teaching and making complex things easy to understand. She has taught numerous courses in Python, R and SQL as a senior data scientist at Metis and as a co-founder of Best Fit Analytics.

She writes about analytics and pop culture on her blog, A Dash of Data. Her work has been featured in Huffington Post, Thrillist and Working Mother. Her highly rated YouTube channel includes popular tutorials such as using data science to settle the age-old cupcake versus muffin debate.

She has spoken at a variety of data science conferences including Strata in New York City and ODSC in San Francisco on topics ranging from natural language processing to data visualization. She has her MS in Analytics and BS in Electrical Engineering, both from Northwestern University.

Carla Gentry

I entered the University of Tennessee (UTC) at Chattanooga in the spring of 1993. I worked in the Developmental Math Lab for my entire tenure with UTC, assisting students with all levels of mathematics. Upon graduating from UTC with a double major, Applied Mathematics and Economics in 1998, I moved to the Chicago area to start my career in analytics.

During that time I worked with Fortune 100 and 500 companies, including: Discover Financial Services, J&J, Hershey, Kraft, Kellogg's, SCJ, McNeil and Firestone. I acted as a liaison between the IT department and the executive staff, and was able to take huge complicated databases, decipher business needs and come back with intelligence that quantifies spending, profit and trends.

I specialize in:

- Comprehensive customer satisfaction and retention analysis

- Brand research and competitive analysis

- Employee retention research

- Survey creation and analysis (new product and branding)

- Database creation and mining

- Social media and coupon incentive promotions

- Project Management (Scrum certified)

Cathy O'Neil

I earned a Ph.D. in math from Harvard, was a postdoc at the MIT math department and a professor at Barnard College. I published a number of research papers in arithmetic algebraic geometry.

I then switched to the private sector, working as a quant for the hedge fund D.E. Shaw in the middle of the credit crisis. I then worked for RiskMetrics, a risk software company that assesses risk for the holdings of hedge funds and banks.

I left finance in 2011 and started working as a data scientist in the New York startup scene, building models that predicted people's purchases and clicks.

I wrote *Doing Data Science* in 2013 and launched the Lede Program in Data Journalism at Columbia University in 2014.

I am a regular contributor to Bloomberg View and wrote the book *Weapons of Math Destruction: How big data increases inequality and threatens democracy.*

I recently founded ORCAA, an algorithmic auditing company.

Claudia Perlich

I started my career in data science at the IBM T.J. Watson Research Center, concentrating on research in data analytics and machine learning for complex real-world domains and applications.

I tend to be domain agnostic, having worked on almost anything from Twitter to DNA, server logs, CRM data, web usage, breast cancer, movie rating and many more. I am still actively publishing.

My claim to fame was winning data-mining competitions, as well as several awards from industry and academia. I received my Ph.D. in Information Systems from Stern School of Business, New York University in 2005 and hold a Master of Computer Science from Colorado University.

Besides my day job, I like teaching and am available for panels, presentations, meetups and debates around big (and small) data.

Deborah Berebichez

I am a physicist, data scientist and television host. I am the first Mexican woman to graduate with a physics Ph.D. from Stanford University.

I co-host Discovery Channel's "Outrageous Acts of Science" TV show (2012 – present) and co-star on the television show "Humanly Impossible" (2011) produced by the National Geographic Channel.

I am chief data scientist at Metis and lead the creation and growth of exceptional data science training opportunities such as boot camps, corporate training and professional development as well as live online programs. In addition, I developed a world-class data science instructional team.

For my doctoral dissertation, I invented a highly effective technique in the field of wireless communications whereby a cell phone user can communicate with a desired target user in a location far away.

My passion is to inspire young people to pursue careers in science. I am an IF/THEN Ambassador for the American Association for the Advancement of Science AAAS, whose goal is to inspire middle and high school girls to pursue careers in STEM.

My STEM outreach has included places such as India, Israel, Mexico and Costa Rica. I have been named a John C. Whitehead Fellow at the Foreign Policy Association and have received the Society of Hispanic Professional Engineers (SHPE)'s Community Service STAR Award. I was also named Top Latina Tech Blogger by the Association of Latinos in Social Media LATISM.

Heather Shapiro

I lead the Data Science function at Pear Therapeutics, which drives:

- Advanced analytics systems and process

- Data-informed decision making across the organization

- Predictive modeling

- Data-driven product and clinical development

- Building the world's first Food and Drug Administration-cleared digital prescription therapies

I studied at the University of California, Davis, and received a Ph.D. in neuroscience in 2013. I have more than 10 years of experience in research and data.

I am passionate about startups, health, wellness, wearables, education and the brain.

Jacqueline Nolis

I am a data science consultant with more than a decade of experience helping companies with data. I have led projects and teams at companies including DSW, Union Bank, and Microsoft.

Prior, I was the director of Insights and Analytics at Lenati and a lead of Advanced Analytics at Promontory Financial Group. I have a Ph.D. in industrial engineering and am also a coauthor of the book *Build Your Career in Data Science*.

I have over a decade of experience in using data science to solve business problems. I've led data science, machine learning, and AI projects at Microsoft, Adobe, Airbnb and T-Mobile and have helped build data science teams from scratch.

Natalie Evans Harris

I speak about the ethical and responsible use of data based on 20 years of advancing the public sector's strategic use of data, including a 16-

year career at the National Security Agency and 18 months with the Obama administration.

I work with a broad network of academic institutions, data science organizations, application developers and foundations to advance the responsible use of data standards, APIs and ethical algorithms to directly benefit people.

I co-founded and currently serve as head of Strategic Initiatives of BrightHive, a data trust platform delivering a suite of smart data collection, integration and governance products to social services providers for improved access to and usability of social sector data.

I founded the community-driven Principles for Ethical Data Sharing (CPEDS) community of practice. It has more than 1,200 active members focused on strengthening ethical practices in the data science community through crowd- sourcing of a Data Science Code of Ethics.

For the Global Data Ethics Project (GDEP), I serve as a strategic adviser in advancing the adoption of tools, techniques and practices developed by the community.

As a senior policy adviser to the U.S. chief technology officer in the Obama administration, I founded The Data Cabinet, a federal data science community of practice with more than 200 active members across more than 40 federal agencies.

I have a master's degree in Public Administration from George Washington University, a Bachelor of Science in Computer Science and a Bachelor of Science in Sociology from University of Maryland Eastern Shore.

Olivia Parr-Rud

I have a Bachelor of Arts in Mathematics and a Masters of Science in Decision and Information Systems with a concentration in Statistics.

For 29 years, I focused on consulting, teaching, and writing about data science with a specialty in predictive modeling for clients such as Cisco, Citizens Bank, Clorox, HP, IBM, and State Farm.

I am a best-selling and award-winning author of five books including, Data Mining Cookbook (Wiley 2001) and Business Intelligence Success Factors (Wiley 2009).

I am now the Go-To Expert for LOVE AT WORK as I champion the power of compassion and caring to drive profits.

ABOUT THE AUTHORS

KATE STRACHNYI'S STORY

I'm the founder of Story by Data and the DATAcated Academy. I'm also just an all-around data science enthusiast. I'm a mother of two little girls (ages 3 and 5).

My children are truly the reason that I ended up working in data science. When I was pregnant with my first daughter, I was working full time serving financial services clients in the regulatory compliance and risk management space. This included heavy travel and long hours (nights and weekends included). I realized that this would not be sustainable and would be extremely difficult to manage once my child was here. Plus, I didn't want to miss seeing her grow up.

This motivated me to speak up in my organization and begin my search for a "remote role," meaning I wanted to work from home (at least part time) to make this transition more manageable. Through persistence and several conversations with leadership, I ended up with a role that involved working with data, and it kept me home about 90% of the time.

I was beyond excited and felt truly grateful for this opportunity. I did everything in my power to ensure I would keep this role. I signed up for data science courses online, watched YouTube tutorials while pumping, and read every article I could get my hands on during maternity leave and during my first few months in the role.

It was around this time (2014) when I realized that I am fascinated by the data science field, so I fully jumped into it. I wrote several articles and recorded countless videos as part of my endeavor to learn, as well as to give back to the data science community.

My projects include Humans of Data Science (a video podcast), DATAcated Challenge (a fun way to learn while collaborating, with a potential to win prizes), *Journey to Data Scientist* (a book where I interviewed more than 20 data scientists on how they ended up in this role), *The Disruptors: Data Science Leaders* (a book that includes an interview and biography of 10 well-known data scientists), and the *Data Literacy for Kids* book.

I remember the exact moment that I decided to write *Mothers of Data Science*. I had taken both of my daughters to the park and was thinking how truly lucky I was to have the flexibility to take them to the park on a workday afternoon. Most of the mothers I met at the park at this time were unemployed (some by choice, some because they had to stay with the kids).

Most of the other kids there were with nannies. I had this thought in the moment: *I wish that more women knew of the opportunities for careers in data science.*

Later in the evening, I messaged Kristen Kehrer on Facebook to ask her if she wanted to co-author a book with me. She said yes without hesitation.

I had met Kristen on LinkedIn about a year before this. I actually interviewed her for the *Humans of Data Science* video podcast, and we realized we had a lot in common. We both knew that the market for the book was not huge, and we were both OK with this. We hope that the stories you've read here have entertained and inspired you.

Kristen Kehrer

I'm Founder of Data Moves Me, LLC, and like Kate said, we have a lot in common. I also have two children (ages 2 and 5) and I love data science. I have a Bachelor of Science in Mathematics and a Master of Science in Applied Statistics. I was working in advanced analytics before the term "data science" was coined, but once the term became popular, I knew that this is what I was, a data scientist.

Throughout my career I haven't been required to travel and I've typically had very comfortable working arrangements. There have been times where I've had to work a weekend or an evening, but for the most part, 40 hours was the standard expectation and it was typically OK if I wasn't there exactly at 9 a.m. or if I left a little before 5.

My 10 years in corporate settings spanned the utility, healthcare, and e-commerce industries, with five years spent in e-commerce.

Nowadays, I help data science teams better communicate machine learning models, insights and recommendations to non-technical stakeholders and the business. I also teach Practical Data Science for the University of California Berkeley Extension, and I'm quite passionate about blogging on different topics in data science on my website at datamovesme.com

Having my own business allows me to get my daughter on and off the school bus and work in close alignment with the school schedule, although my kids will be in camp during the summer.

I had been pursuing remote work for three years before I decided to start building a business on the side that would allow me to be fully remote. I was offered remote full time positions in data science, but they all paid significantly less than my salary at the time, and I wasn't willing to settle for less. So I started my own company, and as a bonus I now also have the flexibility I desired.

Writing this book was such a fun opportunity to speak with other mom data scientists and reflect on my own career journey. Hopefully you'll be inspired to design the data life that works perfectly for you and your needs, wants and desires.

INTERVIEW QUESTIONS

Below are some questions that we asked during the interviews with the mothers of data science.

— Kate & Kristen

1. What are your hopes for your children?

2. What strengths do you have? Did you cultivate this as being a mom?

3. How do you balance continuing education while being a mom?

4. What are some unique strengths in the data science field?

5. Do you focus on teaching your kids data literacy more than you think a non-data mom would?

6. Do you secretly hope your kids will go into a STEM field?

7. Had you thought about staying home with your kids?

8. What are your thoughts on having it all?

9. Is work-life balance a myth? Have you achieved it? Does it get easier as your kids get older?

10. Have you considered stepping down from a high-impact career?

11. How do you manage when your kids are sick and need to be home from school or school concerts, etc.? Is it stressful? Do you feel as though your career uniquely allows you to have flexible time to tend to these situations?

12. Have you ever decided not to go up for promotion to avoid more responsibility?

13. Have you chosen not to travel to avoid being away from kids or turned down management positions?

14. How was your maternity leave experience?

15. How was your experience returning to work?

16. Do you think dads don't think about this as much?

17. Do you have mama's guilt when going to data conferences, traveling, etc.?

18. Have you experienced any events in your data career that you hope your children avoid?

19. Do you work remotely, and do you think it's easier to work remotely as a data scientist?

SURVEY QUESTIONS

These questions were used for our Parents in Data Science survey. Responses were anonymous.

— Kate & Kristen

1. **How would you describe your current role?**

    a. Data Analyst, Machine Learning Engineer, Data Scientist, Consultant, etc.

    b. Manager (or higher) of Data Teams

    c. Trying to get a role in the data field

    d. Other (non data field)

1. **Do you have at least one child living in your home for at least some portion of your work week?**

    a) Yes

    b) No

    c) I used to, but my kid(s) are grown up

2. **How long have you been in the data field?**

    a) Less than 2 years

    b) 2-5 years

    c) 6-10 years

    d) 11 - 15 years

    e) 16 years +

    f) I am not yet in the field

3. **To which gender identity do you most identify?**

    a) Male

    b) Female

    c) Gender variant/Non-conforming

    d) Other

4. **Which category describes your age?**

    a) Younger than 18

    b) 18 - 24

    c) 25 - 34

    d) 35 - 44

    e) 45 - 54

    f) 55 - 64

    g) 65 or older

    h) Prefer not to answer

5. **What best describes your employment status?**

    i) Employed full time

    j) Employed part time

    k) Self-employed

    l) Not employed, but looking for work

    m) Not employed and not looking for work

    n) Retired

    o) Student

    p) Military

    q) Homemaker

    r) Prefer not to answer

6. **Which category best describes your annual income? (in USD)**

    a) Under $29,999

    b) $30,000 - $59,999

    c) $60,000 - $89,999

    d) $90,000 - $119,999

    e) $120,000 - $159,999

    f) $160,000 or more

    g) Prefer not to answer

7. **What type of parental leave did you receive?**

    a) Fully paid

    b) Not paid

    c) Partially paid

    d) Other

8. **How long was your parental leave? Choose the longest leave period if you've had multiple.**

    a) No leave

    b) 2 weeks or less

    c) 3 - 5 weeks

    d) 6 - 8 weeks

    e) 9 - 12 weeks

    f) 13 - 16 weeks

    g) 17 weeks to 6 months

    h) More than 6 months

9. **Where were you located during your parental leave?**

    a) United States

    b) Europe

    c) Asia

    d) Other

10. Do you wish you had a longer parental leave?

a) Yes

b) No

c) Not sure

d) Other

11. Have you ever declined a promotion at work or formally reduced your working hours due to reasons directly related to your child(ren)?

a) Yes

b) No

c) Other

12. Have you ever declined a work trip or to speak at a conference (that would require travel) due to not wanting to leave your child(ren)?

a) Yes

b) No

c) I have not been presented with travel/conference opportunities

d) Other

13. **Have you ever found voicing your needs for time off due to child-related activities difficult (doctor's appointments, school concerts, etc.)?**

    a) Yes

    b) No

    c) Other

14. **Did you feel as though you had a network of people who could help if you needed to stay late at work or needed to transport child(ren) somewhere (other than a spouse)? Family, friends, hired babysitters, etc.**

    a) Yes

    b) No

    c) Other

15. **In general, do you feel as though your partner contributes their fair share in the home (50/50)?**

    a) We equally handle kid-related tasks

    b) My partner handles more of the tasks

    c) My partner handles less of the tasks

    d) I do not have a partner

16. **Have you ever felt you were unfairly treated at work due to having competing priorities related to your children?**

    a) Yes

    b) No

    c) Other

17. **Have you ever changed companies to pursue better work-life balance after having children?**

    a) Yes

    b) No

    c) Other

19. **Do you work at home or in an office?**

    a) Fully remote

    b) Partially remote

    c) Fully in the office

    d) Client's location/Consultant

    e) Other

18. **Want to share any thoughts about juggling work and children? Difficulties, wins or otherwise?**

ABOUT STORY BY DATA INC

Story by Data is "datacated" to providing insights from data. (Data-cated is a play on words of being dedicated to data.)

Story by Data provides services around LinkedIn content strategy for companies focused on innovation in artificial intelligence (AI), machine learning (ML), and data science.

The company is focused on building a data community by hosting the Humans of *Data Science* video podcast series as well as the DATAcated Challenge, a project dedicated to helping others learn about various topics in the data realm (crowd-sourced knowledge).

Story by Data provides online and in-person training on data visualization *via* the DATAcated Academy.

To learn more go to storybydata.com and datacatedacademy.com

ABOUT DATA MOVES ME LLC

Data Moves Me focuses on machine learning storytelling and career development for data scientists. Teaching data scientists to interpret their machine learning models, fully communicate the caveats, influential variables, and insights and recommendations through storytelling and painting a vivid picture in a way that their non-technical stakeholders will understand.

In addition to focusing on machine learning storytelling and career development, Data Moves Me offers a resume course that has helped hundreds of data scientists land phone screens that lead to jobs for data science roles. In addition, Kristen has created a SQL for data science course that is available for free on her website/YouTube.

To learn more go to datamovesme.com

Thank you!

We are really grateful for you taking the time to read the *Mothers of Data Science* book. We hope you found the stories inspirational and relatable. Please feel free to reach out to provide feedback, or to simply connect and have a discussion.

A special thanks to Jennifer Cooper, Tracey DiLascio-Martinuk, Jocelyn Griser, and Tonya Vinas for reviewing the book and providing insightful feedback.

76

— ***Kate Strachnyi*** (*kate@storybydata.com*) and

Kristen Kehrer (*kristen@datamovesme.com*)